Effective Project Management in Offshore Energy Projects

Table of Contents

Preface

Introduction

Prelude

Chapter 1: Key Components of Offshore Energy Projects

Chapter 2: Project Planning and Initiation

Chapter 3: Risk Management and Mitigation Strategies

Chapter 4: Scheduling and Time Management

Chapter 5: Budgeting and Cost Control

Chapter 6: Team Management and Communication

Chapter 7: Quality Management in Offshore Projects

Chapter 8: Safety and Environmental Management

Chapter 9: Project Monitoring and Reporting

Conclusion

Glossary

Preface

The offshore energy industry is at a pivotal point, driven by rapid advancements in technology, evolving regulatory demands, and an increasing emphasis on operational efficiency and sustainability. Managing complex projects in such a dynamic environment requires specialized knowledge and practical tools. Recognizing the need for accessible, targeted information, the *Gosships Learning Series* was developed to provide industry professionals with the expertise they need to manage offshore energy projects effectively.

This series offers foundational to intermediate knowledge with a focus on real-world applications, equipping readers with the skills to execute projects efficiently while ensuring safety and compliance. Each book in the series is accompanied by a certification test to ensure that the knowledge gained is both thoroughly understood and ready to be applied in professional contexts.

The *Gosships Learning Series* aims to empower personnel across the offshore energy sector, from project engineers to site managers, by providing them with essential tools to navigate the complexities of modern offshore energy projects. We hope this series will contribute to your professional growth and help unlock new opportunities for success in your career.

Introduction

Welcome to the *Gosships Learning Series*, designed for professionals who are seeking to expand their knowledge and enhance their project management skills in the offshore energy sector. This book, *Effective Project Management in Offshore Energy Projects*, has been carefully crafted by experienced industry executives and project managers to ensure that the content is both authoritative and aligned with the current best practices in offshore project execution. Whether you're new to the field or seeking to refine your expertise, this resource is tailored to provide insights that will help you lead offshore projects successfully.

In this book, we will explore the following key areas:

- **Project Planning and Scheduling**: Learn the fundamentals of planning and scheduling offshore energy projects, including the use of specialized tools and software for managing complex timelines.

- **Risk Management**: Understand how to identify, assess, and mitigate risks specific to offshore environments, ensuring that your project stays on track.

- **Stakeholder Management**: Explore techniques for managing communication and expectations among key stakeholders, including regulators, contractors, and investors.

- **Health, Safety, and Environmental (HSE) Management**: Delve into the best practices for ensuring the safety of personnel and compliance with environmental regulations throughout the project lifecycle.

- **Cost Control and Budgeting**: Discover strategies for managing project costs, controlling budgets, and preventing cost overruns in high-stakes offshore projects.

Upon completing this book, you will be prepared to take an assessment designed to evaluate your understanding of effective project management in offshore energy settings. By passing the assessment, you will earn a Certificate of Achievement, which can be obtained through www.gosships.com. This certification will validate your skills and demonstrate your expertise to colleagues, stakeholders, and future employers.

Who Is This Book For?

This book is designed for a wide range of professionals involved in offshore energy projects, including:

- **Project Managers and Engineers**: Individuals responsible for planning, executing, and managing offshore energy projects.

- **Shoreside and Offshore Supervisors**: Supervisors overseeing the operational aspects of offshore projects, ensuring that safety, quality, and timelines are maintained.

- **Aspiring Students**: Individuals seeking to enter the offshore energy industry with a strong foundation in project management principles.

- **Government and Regulatory Officials**: Professionals involved in overseeing the compliance and safety standards of offshore energy projects.

By mastering the concepts in this book, you will be better equipped to meet the challenges of managing offshore energy projects, ensuring they are completed on time, within budget, and in compliance with international standards.

Thank you for choosing the *Gosships Learning Series* to support your ongoing journey of professional development and growth in the offshore energy sector.

Gosships Learning Series 2024/2025

1. Hydrogen: The Fuel of the Future
2. Green Ammonia: The Next Big Thing in Shipping
3. Decarbonizing Shipping: Pathways to Zero Emissions
4. Battery Technology for Industrial Applications
5. Carbon Capture and Storage: Can It Save the Planet?
6. Biofuels 101: Turning Waste into Energy
7. Understanding LNG (Liquefied Natural Gas)
8. Methanol as a Marine Fuel
9. Offshore Wind Energy: The Future of Renewable Power
10. Tidal and Wave Energy: Harnessing the Ocean
11. Electrofuels: The Next Generation of Carbon-Neutral Fuels
12. Energy Storage Systems for Grid Reliability
13. Hydrogen Fuel Cells for Transportation
14. Solar Energy Innovations: Beyond Solar Panels
15. Smart Grids: The Backbone of Future Energy Systems
16. Ammonia-Hydrogen Blends: A Dual Fuel Solution?
17. Nuclear Power: Small Modular Reactors for a Low-Carbon Future
18. Hydropower: The Oldest Renewable Energy Source
19. Decentralized Energy Systems: Microgrids for Resilience
20. Energy Efficiency Technologies for Industry
21. Hydrogen Production from Seawater
22. Fuel Cells for Maritime Applications
23. Geothermal Energy: Unlocking Earth's Heat
24. Future of EV Charging Infrastructure
25. Synthetic Fuels: Bridging the Gap to Decarbonization
26. Cybersecurity for Maritime and Offshore Operations
27. AI and Automation in Shipping and Logistics
28. Digital Twins in Maritime: Revolutionizing Asset Management

29	Risk Management in Offshore and Maritime Operations
30	Compliance with IMO 2020 Regulations
31	Sustainable Ship Design: Reducing Environmental Impact
32	Marine Renewable Energy: Wave, Tidal, and Offshore Wind Integration
33	Ballast Water Management Systems
34	Blockchain Technology in Shipping: Improving Transpc'y & Efficiency
35	Effective Supply Chain Management for Energy Industries
36	Leadership in the Energy Transition
37	Effective Crisis Management in Maritime Operations
38	Shipyard Safety Management Systems
39	Port State Control (PSC) Inspection Readiness
40	Remote Vessel Operations and Autonomous Shipping
41	Optimizing Fleet Performance with Data Analytics
42	Maritime Environmental Regulations: Staying Ahead of Compliance
43	Advanced Maintenance Strategies: Condition Monitoring & Predictive Maintenance
44	Global LNG Market: Trends and Opportunities
45	Incident Investigation in Maritime Operations
46	International Maritime Law: Key Concepts and Applications
47	Emergency Preparedness and Response for Offshore Oil & Gas
48	Energy Transition Strategies for Oil and Gas Companies
49	Maritime Drones: Applications and Safety Considerations
50	Effective Project Management in Offshore Energy Projects

All Rights Reserved Disclaimer

The contents of this book, including but not limited to all text, graphics, images, logos, and designs, are the intellectual property of Gosships LLC and are protected by copyright law. No part of this publication may be reproduced, distributed, transmitted, displayed, or modified in any form or by any means, including photocopying, recording, or other electronic or mechanical methods, without the prior written permission of the publisher, except in the case of brief quotations in critical reviews or articles.

The information contained within this book is for educational purposes only and is provided "as is" without warranty of any kind, either expressed or implied. The authors and publishers disclaim any liability for any direct, indirect, or consequential loss or damage arising from the use of the material in this book.

For permissions or inquiries, please contact: admin@gosships.com

© 2024 Gosships LLC. All rights reserved.

Prelude

Offshore energy projects are central to the global energy sector, whether they involve oil and gas extraction or renewable energy sources like offshore wind and tidal power. These projects require meticulous planning and execution due to the unique challenges posed by the marine environment. Managing such projects effectively ensures not only timely and on-budget delivery but also the health and safety of personnel, the protection of the environment, and adherence to strict regulatory frameworks.

The intricacies of offshore energy projects—ranging from logistical issues, equipment procurement, extreme weather conditions, and high stakeholder involvement—necessitate a robust project management framework. This mini-book provides a thorough exploration of the key components, tools, and strategies essential for managing offshore energy projects. By focusing on risk management, scheduling, budgeting, and team coordination, this guide aims to provide readers with a solid foundation for managing complex offshore endeavors.

Chapter 1

Key Components of Offshore Energy Projects

1.1 Project Scope

Defining the project scope is a critical first step in managing offshore energy projects. Offshore projects differ from onshore projects due to their location, complexity, and scale. Understanding the full breadth of work required, from exploratory surveys to decommissioning, is essential. Offshore energy projects often include:

1. **Exploration**: The exploration phase involves identifying viable offshore resources, such as oil, gas, or wind potential. This includes geological and seismic surveys, environmental impact assessments (EIAs), and determining the feasibility of the project. During this phase, teams assess risks, regulatory requirements, and environmental considerations.

2. **Construction and Installation**: This phase involves building and installing the necessary infrastructure. For oil and gas, this includes drilling rigs, production platforms, pipelines, and transportation infrastructure. For offshore wind, it includes turbine foundations, cables, and substations. The complexity of constructing in harsh offshore conditions, combined with the logistical challenges of transporting materials, makes this phase particularly demanding.

3. **Operations and Maintenance**: After construction, the project moves into the operations phase. This involves managing the extraction or production of energy, routine inspections, maintenance, and continuous monitoring of equipment performance. Maintenance in offshore projects is particularly challenging due to limited access and extreme weather conditions, requiring highly specialized personnel.

4. **Decommissioning**: The end-of-life phase for offshore energy projects involves dismantling and safely removing the infrastructure. This phase must be meticulously planned to minimize environmental impacts and adhere to local regulations regarding the safe disposal of equipment and waste.

1.2 Stakeholder Engagement

Offshore energy projects involve a wide array of stakeholders, each with different expectations, goals, and concerns. The success of the project hinges on the project manager's ability to effectively engage with these stakeholders. Key stakeholders typically include:

- **Investors and Financiers**: These stakeholders provide the capital required for the project and are keenly interested in the project's return on investment (ROI). Maintaining financial transparency and providing regular updates on project milestones are essential to securing ongoing support.

- **Regulatory Bodies and Governments**: Offshore projects are heavily regulated, particularly with regard to environmental protection, worker safety, and maritime law. Agencies such as the Environmental Protection Agency (EPA) and the International Maritime Organization (IMO) set strict guidelines that must be followed. Maintaining compliance and securing the necessary permits and licenses is a time-consuming yet essential process.

- **Local Communities**: Coastal and indigenous communities may be impacted by offshore projects, especially in terms of environmental changes and job creation. Engaging with local communities, addressing concerns about environmental impacts, and offering employment opportunities help ensure a positive relationship with these stakeholders.

- **Internal Teams and Contractors**: Offshore energy projects are multidisciplinary and often involve international teams. Managing communication and collaboration between engineers, environmental scientists, construction crews, and administrative teams is crucial to ensure the project runs smoothly.

Chapter 2
Project Planning and Initiation

2.1 Setting Clear Objectives

Setting well-defined objectives is essential for any project, but it is particularly important in the offshore energy sector due to the high costs and risks involved. Objectives must be:

- **Specific**: Objectives should be detailed and specific enough to guide every phase of the project. For example, instead of "install wind turbines," a specific objective would be "install 25 5MW wind turbines by Q3 of the project year."

- **Measurable**: It's crucial to have measurable metrics to track progress and performance. Key metrics might include the number of completed wells, percentage of turbines installed, or energy output targets.

- **Achievable**: The objectives must be realistic given the project's budget, resources, and timeline. Overly ambitious targets can strain resources, causing delays and financial overruns.

- **Relevant**: Objectives must align with the broader goals of the company and the stakeholders. For example, a project focusing on renewable energy should have objectives that reflect sustainability and long-term viability.

- **Time-bound**: Setting deadlines is critical to ensure the project stays on track. Offshore projects often face delays due to weather or logistical challenges, so timelines should account for these potential disruptions.

2.2 Project Charter and Documentation

A well-crafted project charter serves as the guiding document for the entire project. The charter outlines the project's purpose, its high-level objectives, and the stakeholders involved. It typically includes:

- **Project Vision and Goals**: The overall purpose of the project and what it aims to achieve.

- **Key Deliverables**: Major milestones and outputs expected during the project lifecycle.

- **Budget Overview**: A high-level view of the total project budget, including contingency plans for overruns.
- **Team Structure**: Identifying the core project team, their roles, and lines of reporting.
- **Risk Management Strategy**: Outlining initial risks identified during the project planning phase.

In addition to the project charter, maintaining detailed documentation throughout the project is essential for transparency, communication, and compliance. This includes maintaining up-to-date risk registers, project schedules, and procurement logs.

2.3 Resource Allocation

Offshore projects require significant resources, including personnel, specialized equipment, and technology. Efficiently allocating these resources is critical to meeting project deadlines and staying within budget. Factors that affect resource allocation include:

- **Personnel**: Ensuring you have the right mix of skills on board, from marine engineers to environmental experts, is critical. Offshore teams often operate in remote locations, so access to medical staff and safety officers is also essential.
- **Equipment**: Construction and installation phases may require heavy-duty cranes, vessels, helicopters, and other specialized equipment. Delays in procuring or deploying equipment can cause significant delays in the project schedule.
- **Logistics**: Transporting personnel, equipment, and materials to and from offshore sites involves complex logistical planning. Delays in logistics can be caused by adverse weather, transport bottlenecks, or regulatory issues. Planning for redundancies and having contingency options in place is vital.

Chapter 3
Risk Management and Mitigation Strategies

3.1 Identifying Risks

Offshore energy projects face a myriad of risks that can result in delays, cost overruns, or even project failure. Common risks include:

- **Technical Risks**: Offshore equipment is exposed to harsh conditions such as saltwater corrosion, high winds, and deepwater pressures. These factors can cause equipment failures, leaks, and other issues that halt progress.

- **Operational Risks**: These include crew safety, labor shortages, and coordination difficulties between offshore and onshore teams. Safety regulations and protocols must be strictly followed to prevent accidents.

- **Environmental Risks**: Extreme weather events like hurricanes or storms pose significant risks to offshore energy projects. Additionally, environmental accidents, such as oil spills or turbine collapses, can result in severe damage to marine ecosystems and substantial financial penalties.

- **Financial Risks**: Cost overruns and changes in global energy prices can affect the profitability of the project. Offshore projects often involve volatile materials markets, where steel prices, for example, may fluctuate significantly.

3.2 Risk Assessment Tools

There are several tools used to assess and categorize risks in offshore energy projects:

- **Risk Matrix**: This tool allows project managers to assess risks based on their likelihood and impact. High-impact, high-likelihood risks are prioritized for mitigation.

- **Failure Mode and Effects Analysis (FMEA)**: A systematic approach for identifying potential failure points in the project's processes and assessing their impact.

- **SWOT Analysis (Strengths, Weaknesses, Opportunities, Threats)**: This is particularly useful in the planning stage for identifying internal and external risks.

3.3 Mitigation Planning

Mitigating risks involves developing strategies to either reduce the likelihood of the risk occurring or minimize its impact if it does occur. Common mitigation strategies include:

- **Contingency Plans**: Building extra time into the project schedule to accommodate potential delays, especially those caused by weather or equipment failures.

- **Redundant Systems**: Installing backup systems and infrastructure, such as additional power sources, to ensure operations can continue if a primary system fails.

- **Insurance and Contract Clauses**: Including clauses in contracts to cover delays, accidents, or other unforeseeable issues can protect the project from financial losses.

Chapter 4

Scheduling and Time Management

4.1 Project Scheduling Techniques

Effective scheduling is essential for delivering offshore energy projects on time. Some common scheduling techniques include:

- **Gantt Charts**: Gantt charts visually represent the project timeline, showing each task, its duration, and how it fits into the overall schedule. This tool helps teams understand dependencies between tasks and adjust timelines when necessary.

- **Critical Path Method (CPM)**: CPM is used to identify the sequence of tasks that must be completed on time for the entire project to stay on schedule. Delays in tasks on the critical path can cause the whole project to fall behind.

- **Program Evaluation Review Technique (PERT)**: PERT is a statistical tool that helps project managers estimate the time required for each task. By incorporating optimistic, pessimistic, and most likely time estimates, PERT accounts for uncertainty in task duration.

4.2 Managing Delays

Delays are a common challenge in offshore projects, and managers must have strategies in place to minimize their impact. Common causes of delays include:

- **Weather Conditions**: High winds, storms, and rough seas can halt operations, particularly during the installation phase. Project managers can mitigate weather-related delays by scheduling work during favorable weather seasons or building weather contingencies into the schedule.

- **Equipment Issues**: Malfunctions, breakdowns, or delays in equipment delivery can severely impact timelines. Ensuring that equipment is properly maintained and scheduling regular inspections can help mitigate these risks.

- **Regulatory Delays**: Permitting and regulatory approvals can take longer than expected, particularly in sensitive environmental zones. Early engagement with regulatory bodies and thorough documentation can help avoid delays.

4.3 Time Tracking and Adjustments

Time tracking tools help managers monitor progress and make necessary adjustments. Software such as **Primavera P6** or **Microsoft Project** allows project managers to update timelines in real-time, making it easier to identify delays and adjust resources accordingly. Regular review meetings ensure that the project stays on track, and any emerging issues are dealt with swiftly.

Chapter 5
Budgeting and Cost Control

5.1 Project Budgeting Essentials

Offshore energy projects are capital-intensive, requiring precise budgeting and cost management. The budget must account for:

- **Exploration and Surveys**: This includes geological and seismic studies, environmental impact assessments, and legal permits. These costs are incurred before the project even begins construction and can vary based on the region and regulatory environment.

- **Construction Costs**: Major expenses in this phase include materials (steel for platforms or turbines), labor, transport, and installation equipment. The cost of transporting materials and personnel offshore can be substantial due to the remote nature of these projects.

- **Operational Costs**: These include the ongoing costs of running the project, such as personnel salaries, energy to run equipment, and regular maintenance. In offshore wind farms, operational costs also include the cost of grid connection and energy distribution infrastructure.

- **Decommissioning Costs**: At the end of the project's life, the infrastructure must be safely dismantled and removed. Decommissioning must be planned early to ensure that funds are available, and costs are controlled.

5.2 Cost Estimation Techniques

Accurate cost estimation is vital for securing investment and ensuring the project stays within budget. Techniques include:

- **Bottom-Up Estimating**: This method involves estimating the cost of each individual task or component of the project and then aggregating those estimates to arrive at the total cost. This is a highly detailed approach but time-consuming.

- **Analogous Estimating**: This method relies on data from similar projects completed in the past to estimate costs. It's faster but less precise than bottom-up estimating.

- **Parametric Estimating**: This statistical approach uses data from previous projects and applies mathematical formulas to estimate the cost of a new project. For example, it might estimate the cost per kilometer of undersea pipeline based on data from past projects.

5.3 Monitoring Project Costs

Throughout the project, managers must continuously monitor costs to ensure they remain within the budget. **Earned Value Management (EVM)** is a useful technique for comparing the project's budgeted costs against actual spending. By analyzing variances, project managers can determine whether the project is on budget, behind schedule, or at risk of exceeding the budget.

Regular financial reporting is essential to track cash flow, identify cost overruns, and adjust resource allocation as needed. Tools like **SAP** or **Oracle Primavera** help in tracking costs in real-time, enabling managers to make quick adjustments.

Chapter 6
Team Management and Communication

6.1 Building Effective Teams

Offshore energy projects require highly skilled and multidisciplinary teams, including engineers, environmental experts, safety officers, logistics managers, and project managers. An effective project team should:

- **Clearly Define Roles and Responsibilities**: Team members should understand their specific responsibilities and how they contribute to the overall project. For instance, a marine engineer's role differs significantly from that of an environmental compliance officer.

- **Foster Collaboration**: Offshore projects require collaboration between various disciplines. Facilitating communication and fostering a collaborative culture is essential for ensuring that different teams work together toward common goals.

- **Balance Onshore and Offshore Resources**: Offshore teams face unique challenges, including isolation, long working hours, and harsh conditions. Managers must ensure that offshore teams are supported by onshore staff and that rotation schedules are designed to prevent fatigue and burnout.

6.2 Communication Tools

Communication is the lifeblood of any project, but in offshore energy projects, it becomes even more critical due to the remote locations of teams. Key tools for maintaining clear and consistent communication include:

- **Project Management Software**: Platforms like **Asana**, **Trello**, and **Microsoft Teams** help team members stay aligned, track progress, and communicate issues in real time.

- **Video Conferencing and Communication Channels**: Offshore teams can feel isolated from onshore operations, so regular video conferences ensure that everyone is on the same page. Tools like **Zoom** and **Microsoft Teams** enable face-to-face communication, even from remote locations.

- **Daily Briefings and Reporting**: On-site teams, particularly in offshore environments, benefit from daily briefings that outline the day's objectives, review potential risks, and address any issues from the previous day. These briefings improve coordination and help prevent accidents.

6.3 Conflict Resolution

Given the scale and complexity of offshore projects, conflicts between team members or between contractors and the project owner are inevitable. Project managers must be adept at resolving conflicts through:

- **Mediation**: Encouraging open dialogue between conflicting parties to resolve issues before they escalate.
- **Clear Communication Channels**: Establishing clear communication pathways so that concerns can be raised early and dealt with quickly.

6.4 Cultural and Geographical Considerations

Offshore energy projects often involve multinational teams working across different regions. Project managers need to:

- **Acknowledge Cultural Differences**: Recognize that team members may have different working styles, communication preferences, and expectations based on their cultural background. Building cultural awareness into team development fosters cooperation and mutual respect.
- **Manage Time Zones**: Offshore teams frequently operate in different time zones from onshore support teams. Project managers must schedule meetings and communication to accommodate everyone's working hours.

Chapter 7
Quality Management in Offshore Projects

7.1 Quality Control vs. Quality Assurance

Quality management ensures that the project meets its technical and regulatory requirements while minimizing the risk of costly defects. There are two key components of quality management:

- **Quality Control (QC)**: QC focuses on the operational aspects of ensuring that project outputs meet predefined standards. This involves testing materials, inspecting installations, and verifying that construction processes comply with design specifications.

- **Quality Assurance (QA)**: QA is a proactive approach that aims to prevent defects before they occur. This involves setting up processes and standards at the start of the project to ensure that all work meets the necessary quality levels. It includes thorough planning, training, and continuous improvement processes.

7.2 Standards and Compliance

Offshore projects must adhere to stringent international and local standards, including:

- **ISO 9001**: An international standard that specifies the requirements for a quality management system (QMS).

- **API Standards (American Petroleum Institute)**: API standards govern oil and gas equipment and operational practices to ensure safety and environmental protection.

- **DNV GL Standards**: DNV GL provides technical standards and certification for offshore structures, particularly in the oil and gas and renewable energy sectors.

Ensuring compliance with these standards involves regular audits, third-party inspections, and internal reviews to verify that all processes are in line with industry best practices.

7.3 Inspection and Testing

Regular inspections and testing are critical to maintaining quality and safety in offshore projects. Inspection processes typically include:

- **Pre-installation Inspections**: Verifying that all materials, equipment, and infrastructure meet the necessary quality standards before being installed offshore.

- **Operational Testing**: Testing installed systems to ensure they function as designed and can withstand offshore conditions, such as extreme weather or deep-sea pressures.

- **End-of-Life Inspections**: When decommissioning offshore assets, rigorous inspections are conducted to ensure that infrastructure is removed safely, without harming the environment or violating regulatory guidelines.

Chapter 8

Safety and Environmental Management

8.1 Safety Protocols and Regulations

Safety is a top priority in offshore energy projects due to the high risks posed by the environment. The regulatory framework governing safety includes:

- **Occupational Safety and Health Administration (OSHA)**: In the U.S., OSHA sets safety guidelines for workers in offshore environments, covering everything from emergency evacuation procedures to protective gear requirements.

- **International Maritime Organization (IMO)**: IMO regulations govern the safety of offshore installations and maritime transport, ensuring that all operations at sea adhere to global safety standards.

- **Offshore Safety Directive (European Union)**: This directive sets strict safety requirements for offshore oil and gas operations within EU waters, with an emphasis on accident prevention, emergency preparedness, and response protocols.

8.2 Environmental Impact Management

Offshore projects pose significant environmental risks, including potential oil spills, water pollution, habitat destruction, and marine wildlife disruption. To mitigate these risks, environmental management protocols must be strictly followed. These include:

- **Environmental Impact Assessments (EIAs)**: Required for all offshore projects, these assessments evaluate the potential environmental consequences of the project and recommend measures to mitigate negative impacts.

- **Spill Response Plans**: Offshore oil and gas projects, in particular, must have well-developed spill response plans in place, including immediate containment measures and long-term cleanup strategies.

- **Waste Management**: Offshore projects generate significant waste, from drilling fluids to discarded materials. Effective waste management systems, including recycling and safe disposal of hazardous materials, are essential for minimizing environmental impact.

8.3 Sustainability Initiatives

In recent years, sustainability has become a key focus in the offshore energy industry, particularly with the rise of offshore wind and other renewable energy projects. Sustainable practices that can be incorporated into offshore projects include:

- **Reducing Carbon Emissions**: Offshore wind farms have a lower carbon footprint than traditional oil and gas operations. Project managers can further reduce emissions by using energy-efficient equipment and processes.

- **Protecting Marine Life**: Measures such as monitoring the impact of noise pollution on marine mammals or restricting operations during sensitive seasons (e.g., whale migration) can help mitigate harm to marine ecosystems.

- **Using Renewable Materials**: Where possible, using sustainable or recycled materials in construction can help reduce the overall environmental impact of the project.

Chapter 9
Project Monitoring and Reporting

9.1 Performance Metrics

Effective monitoring is essential to track the progress of offshore projects. Key performance indicators (KPIs) typically include:

- **Schedule Performance Index (SPI)**: This KPI measures whether the project is ahead of, on, or behind schedule. It is calculated by comparing the value of work completed to the planned value.

- **Cost Performance Index (CPI)**: CPI compares the cost of work performed to the budgeted cost, helping project managers determine whether the project is on budget.

- **Safety Metrics**: These include tracking the number of accidents, near-misses, and safety violations. Offshore projects must adhere to strict safety protocols, and safety metrics are a key measure of project success.

- **Environmental Metrics**: Monitoring metrics such as emissions, waste produced, and water quality ensures that the project adheres to environmental regulations and sustainability goals.

9.2 Reporting Tools

To ensure transparency and communication with stakeholders, regular reporting is essential. Reporting tools include:

- **Dashboards**: Visual reporting tools that provide an at-a-glance overview of project performance, including cost, schedule, and safety metrics.

- **Project Management Software**: Tools like **Microsoft Project** or **Primavera P6** allow project managers to track progress, generate reports, and update stakeholders regularly.

- **Progress Reports**: These should be issued on a regular basis (e.g., weekly or monthly) and include a summary of project achievements, upcoming milestones, budget updates, and any emerging risks or issues.

9.3 Managing Project Changes

Changes are inevitable in offshore energy projects, whether due to shifting market conditions, regulatory requirements, or unforeseen risks. A well-defined **change management process** helps ensure that changes are implemented smoothly, with minimal disruption to the project's overall timeline and budget. This process includes:

- **Change Requests**: Documenting the reasons for a proposed change, its potential impact on the project, and obtaining approval from stakeholders.

- **Impact Assessment**: Analyzing the financial, operational, and schedule implications of the change before making adjustments to the project plan.

- **Stakeholder Communication**: Keeping all relevant stakeholders informed of changes and ensuring their agreement with any modifications to the project's scope, schedule, or budget.

Conclusion

Offshore energy projects present complex and unique challenges, requiring robust project management practices to ensure success. From the initial exploration phase through to construction, operation, and decommissioning, each stage of the project lifecycle demands careful planning, risk management, and collaboration. By focusing on key areas such as budgeting, scheduling, team management, safety, and environmental protection, project managers can navigate the many obstacles of offshore energy projects and deliver successful outcomes.

This mini-book has provided a comprehensive overview of the essential principles of offshore project management. The rise of renewable energy sources like offshore wind further underscores the importance of skilled project management in the energy transition. By mastering these techniques, project managers will be well-prepared to lead the next generation of offshore energy projects.

Glossary: Effective Project Management in Offshore Energy Projects

1. **Baseline**: A fixed project schedule, budget, or scope that serves as a reference point for measuring project performance and progress.

2. **Budget Overrun**: The amount by which the actual cost of a project exceeds the planned budget.

3. **Change Order**: A document that outlines modifications to the scope, schedule, or cost of a project after it has begun.

4. **Client**: The organization or entity that commissions an offshore energy project, responsible for providing funding and approval.

5. **Commissioning**: The process of testing and verifying that systems, equipment, and processes are functioning as required before the project is fully operational.

6. **Communication Plan**: A strategy for ensuring effective communication between project stakeholders, including updates on project progress, risks, and changes.

7. **Compliance**: Adherence to laws, regulations, standards, and project-specific requirements, particularly in safety and environmental matters.

8. **Contingency Plan**: A plan developed to address potential risks and unexpected issues that may arise during a project, ensuring that disruptions are minimized.

9. **Cost Control**: The process of managing project expenses to keep them within the approved budget.

10. **Critical Path Method (CPM)**: A project management technique that identifies the longest sequence of tasks that must be completed on time for the project to meet its deadline.

11. **Deliverable**: A tangible or intangible output produced as part of a project, such as a report, system, or completed infrastructure.

12. **Earned Value Management (EVM)**: A project management technique used to measure project performance and progress by comparing planned and actual costs and schedules.

13. **Environmental Impact Assessment (EIA)**: A process to evaluate the potential environmental effects of an offshore energy project, ensuring regulatory compliance and sustainability.

14. **Feasibility Study**: An analysis conducted to assess whether a proposed offshore energy project is viable in terms of technical, financial, and environmental factors.

15. **Float (Slack)**: The amount of time a task can be delayed without affecting the project's completion date, important for flexibility in scheduling.

16. **Gantt Chart**: A visual representation of a project's schedule, showing tasks, durations, and dependencies.

17. **Health, Safety, and Environmental (HSE) Plan**: A document outlining the safety procedures, regulations, and environmental considerations to be followed during the project.

18. **Initiation Phase**: The first phase of a project, in which objectives, scope, and feasibility are defined, and key stakeholders are identified.

19. **Integrated Project Management (IPM)**: A holistic approach that ensures all project processes, teams, and functions work together seamlessly for project success.

20. **Key Performance Indicator (KPI)**: A measurable value that demonstrates how effectively a project is achieving its objectives.

21. **Kickoff Meeting**: The first official meeting held to introduce the project team, review the project scope, and establish roles and responsibilities.

22. **Lessons Learned**: Documented experiences and insights gained from the project, used to improve future projects and prevent recurring issues.

23. **Milestone**: A significant event or achievement within a project, often marking the completion of a key phase or deliverable.

24. **Monte Carlo Simulation**: A risk analysis technique used to model the probability of different outcomes in project planning, accounting for uncertainty and variability.

25. **Offshore Installation**: The process of placing infrastructure, such as oil rigs, wind turbines, or pipelines, into the offshore environment.

26. **Operational Readiness**: The state at which a project is fully functional and ready for handover, ensuring all systems are in place for successful operation.

27. **Permitting**: The process of obtaining the necessary legal and regulatory approvals to execute an offshore energy project.

28. **Phase Gate**: A decision point at the end of each project phase where the project's progress is evaluated before proceeding to the next phase.

29. **Project Charter**: A formal document that defines a project's objectives, scope, stakeholders, and authority, typically issued at the start of a project.

30. **Project Execution Plan (PEP)**: A document outlining how the project will be carried out, including timelines, resources, risks, and responsibilities.

31. **Project Lifecycle**: The stages a project goes through from initiation to closure, typically including initiation, planning, execution, monitoring, and closure.

32. **Project Management Office (PMO)**: A centralized team or department that oversees project management practices, standards, and methodologies across an organization.

33. **Project Scope**: The detailed description of the work required for the project, including tasks, deliverables, and objectives.

34. **Project Sponsor**: The individual or group responsible for funding and supporting the project, often representing the interests of the client.

35. **Punch List**: A list of items or tasks that need to be completed or corrected before the project can be officially considered complete.

36. **Quality Assurance (QA)**: The process of ensuring that project deliverables meet defined quality standards through systematic monitoring and evaluation.

37. **Quality Control (QC)**: The operational techniques and procedures used to ensure that deliverables meet the required specifications and standards.

38. **Regulatory Compliance**: Adherence to all laws, regulations, and standards relevant to the offshore energy project, especially those concerning safety, environmental protection, and labor.

39. **Request for Proposal (RFP)**: A document that solicits proposals from vendors or contractors for the work required in the project.

40. **Resource Allocation**: The process of assigning available resources—such as personnel, equipment, and budget—to project tasks to ensure efficient completion.

41. **Risk Mitigation**: Strategies and actions taken to reduce or eliminate the impact of risks on the project's objectives, timeline, or budget.

42. **Safety Case**: A comprehensive document that outlines the safety measures and risk assessments for an offshore energy project, required by regulators to ensure safe operations.

43. **Schedule Baseline**: The approved version of a project's schedule, which serves as the standard against which progress is measured.

44. **Scope Creep**: The uncontrolled expansion of a project's scope without corresponding increases in time, budget, or resources, often leading to delays or cost overruns.

45. **Stakeholder**: Any individual, group, or organization that has an interest in the project and can influence its outcome or be affected by it.

46. **Statement of Work (SOW)**: A document that outlines the tasks, deliverables, timelines, and responsibilities of the parties involved in the project.

47. **Sustainability**: The ability to meet the needs of the present without compromising future generations, particularly in offshore projects focused on energy efficiency and environmental protection.

48. **Task Dependency**: The relationship between project tasks where one task must be completed before another can begin, critical for planning and scheduling.

49. **WBS (Work Breakdown Structure)**: A hierarchical decomposition of the project into smaller, manageable components, defining the project's scope and deliverables.

50. **Work Package**: A unit of work that is part of the WBS, assigned to a team or individual, detailing tasks and responsibilities for project completion.

www.ingramcontent.com/pod-product-compliance
Lightning Source LLC
Chambersburg PA
CBHW030041230526
45472CB00002B/611